HOW IT WORKS

Military Machines

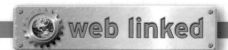

Military Machines

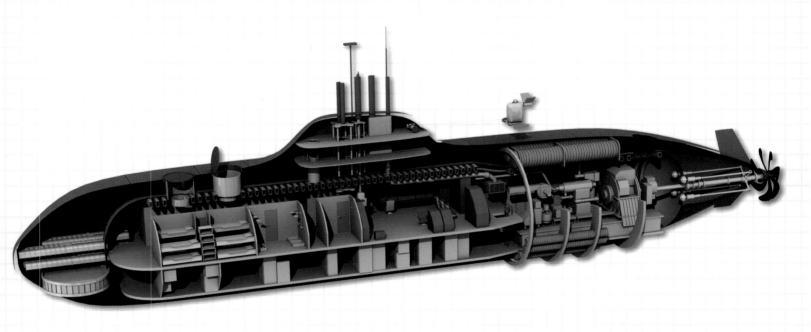

By Steve Parker
Illustrated by Alex Pang

MASON CREST PUBLISHERS INC.
370 Reed Road, Broomall, Pennsylvania 19008
(866)MCP-BOOK (toll free), www.masoncrest.com

First Printing
9 8 7 6 5 4 3 2 1

Library of Congress Cataloging-in-Publication Data
Parker, Steve, 1952–
  Military machines / by Steve Parker ; llustrated by Alex Pang.
    p. cm. — (How it works)
  Originally published: Great Bardfield, Essex : Miles Kelly Pub., c2009.
  Includes bibliographical references and index.
  ISBN 978-1-4222-1797-9
  Series ISBN (10 titles): 978-1-4222-1790-0
  1. Military weapons—Juvenile literature. 2. Vehicles, Military—
Juvenile literature. 3. Airplanes, Military—Juvenile literature.
4. Warships—Juvenile literature. I. Pang, Alex. II. Title.
  UF500.P368 2011
  623.4—dc22
                    2010033619
Printed in the U.S.A.

First published by Miles Kelly Publishing Ltd
Bardfield Centre, Great Bardfield, Essex, CM7 4SL
© 2009 Miles Kelly Publishing Ltd

Editorial Director: *Belinda Gallagher*
Art Director: *Jo Brewer*
Design Concept: *Simon Lee*
Volume Design: *Rocket Design*
Cover Designer: *Simon Lee*
Indexer: *Gill Lee*
Production Manager: *Elizabeth Collins*
Reprographics: *Stephan Davis*
Consultants: *John and Sue Becklake*

Every effort has been made to acknowledge the source and copyright
holder of each picture. The publisher apologizes for any unintentional
errors or omissions.

ACKNOWLEDGEMENTS

All panel artworks by Rocket Design
The publishers would like to thank the following sources
for the use of their photographs:
Alamy: 7(c) Purestock; 30 Peter Jordan; 33 vario images
GmbH & Co.KG
Corbis: 6(t) Bettmann; 6 (b) Corbis; 14 Jeon Heon-Kyun/
epa; 23 Shawn Thew/epa; 26 Pawel Supernak/epa
Getty Images: 13, 18, 20 Time & Life Pictures; 35 Getty
Images
Photolibrary: 9 Larry McManus; 24 Philip Wallick;
36 US. Navy
Rex Features: 7(c); 11; 17 Greg Mathieson
All other photographs are from Miles Kelly Archives

WWW.FACTSFORPROJECTS.COM

CONTENTS

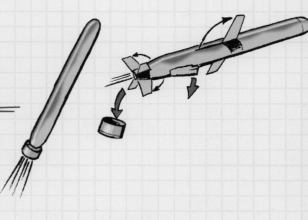

INTRODUCTION. 6

M3M 50-CAL MACHINE GUN. 8

TOMAHAWK AND GRANIT MISSILES . 10

M2 BRADLEY IFV 12

M270 ROCKET LAUNCHER 14

M1 ABRAMS TANK 16

AH-64 APACHE HELICOPTER. 18

A-10 THUNDERBOLT 20

V-22 OSPREY TILTROTOR 22

F-22 RAPTOR 24

E-3 SENTRY AWACS. 26

B-52 STRATOFORTRESS 28

AVENGER MINEHUNTER 30

TYPE 45 DESTROYER 32

TYPE 212 SUBMARINE 34

NIMITZ SUPERCARRIER 36

GLOSSARY. 38

INDEX 40

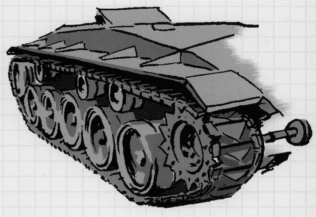

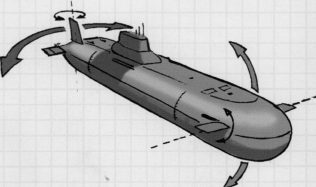

INTRODUCTION

Since history was first recorded, people have waged war on each other. Some want to take power and rule, while others just wish to defend their homelands and citizens. The end result often depends on military technology. Through the ages, new inventions such as gunpowder, cannons, tanks, submarines, bomber planes, and missiles have been the difference between defeat and success. Developments in warfare also have spin-offs for ordinary life, from warmer clothes to the latest electronic gadgets.

The Romans conquered dozens of nations with the latest siege engines and catapults.

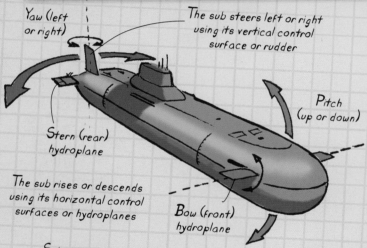

Yaw (left or right)

The sub steers left or right using its vertical control surface or rudder

Stern (rear) hydroplane

Pitch (up or down)

The sub rises or descends using its horizontal control surfaces or hydroplanes

Bow (front) hydroplane

Submarines were the sneaky new weapons of World War II, and were nicknamed 'The Secret Service'.

LET BATTLE BEGIN

Military machines are born from the technology of their time—and then they push it to the limit. In ancient times, wood and stone were naturally the first weapons. The Iron Age brought heavier hammers and sharper blades. Archimedes of ancient Greece invented a massive lever-crane to grab enemy ships and shake them to bits. Medieval scientists came up with catapults and rams to knock down the strongest castle walls.

WORLD AT WAR

In 1914 the first of two World Wars began. Early cars, trucks, and aircraft broke down on a regular basis. However, by the time World War I ended in 1918 there were powerful tanks, huge guns on railway wagons, off-road vehicles, and many kinds of planes including fighters and bombers. The most powerful war machines were battleships. Not many years later, World War II (1939–45) attacks combined huge forces on land, at sea and in the air.

Douglas Dauntless dive-bombers fly over the US aircraft carrier Enterprise, ready for a decisive World War II battle.

The military machines featured in this book are Internet linked.
Visit www.factsforprojects.com to find out more.

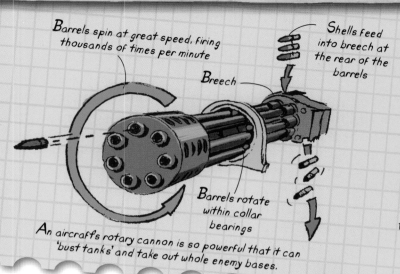

Barrels spin at great speed, firing
thousands of times per minute

Breech

Shells feed
into breech at
the rear of the
barrels

Barrels rotate
within collar
bearings

An aircraft's rotary cannon is so powerful that it can
'bust tanks' and take out whole enemy bases.

TOP GUNS

By the 1950s, air forces had come into their own. One long-range bomber could carry enough nuclear weapons to destroy a whole country. Fighters rode "shotgun" to protect it, while reconnaissance (spy) aircraft went ahead to check the route. On the ground, tanks were faster and better armored, so bigger, more powerful guns were invented to smash them to bits. The first ship-launched missiles appeared— unmanned machines took over from people.

UNSEEN ENEMY

In recent years, it's no longer a show of strength that wins battles. It's no show at all. Stealth is the latest trend. Planes, ships, and ground forces must stay hidden from the enemy's radar, heat-seeking missiles and microwave messages. There is a need for less explosives and more electronics. In the future, military power will continue to change. So what's the next big thing in the arms race?

Ships carrying massive missile firepower can
speed to anywhere in the world.

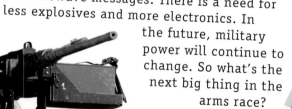

Heavily armed Land Rovers
patrol local conflict in Helmand
Province, Afghanistan.

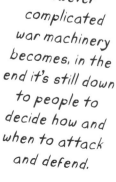

However
complicated
war machinery
becomes, in the
end it's still down
to people to
decide how and
when to attack
and defend.

M3M 50-CAL MACHINE GUN

The Browning M3M is a modern variant of John Browning's tried and tested M2 .50 Caliber heavy machine gun, which came into service more than 90 years ago after World War I (1914–18). Used by forces the world over, the M2 has been the model for many rapid-fire variants since. The M3M fires more than 1,000 rounds per minute—that's 18 bullets each second.

Eureka!

John Browning (1855–1926) was just a boy when he built a rifle from bits of scrap metal. His idea for automatic weapons (weapons that rapidly reloaded themselves as they fired) came from the hand-cranked Gatling gun of the era (see page 20).

What next?

Experiments with high-power laser "ray guns" show they can carry enough energy to explode munitions a third of a mile (half a kilometer) away.

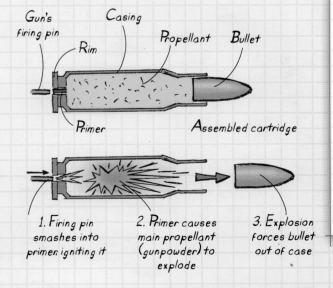

Rear sight

Cocking handle

Front sight

Spade grip

Recoil mechanism Energy from the "kick" or recoil of the fired bullet is used to eject the empty, or spent, cartridge casing and load the next cartridge into position, ready to fire.

Spent cartridge

✳ How do BULLETS work?

Bullets are usually solid metal or plastic and do their damage by smashing into the target. (Shells are similar but contain explosives that blow up when they hit an object.) A typical machine gun bullet is packaged into a unit called a cartridge along with its casing, explosive and primer. The primer is a small amount of explosive that ignites easily when the firing pin bangs into it. In turn, it sets off the main explosive propellant, which blasts the bullet from its casing and along the barrel.

Gun's firing pin

Casing

Rim

Propellant

Bullet

Primer

Assembled cartridge

1. Firing pin smashes into primer igniting it

2. Primer causes main propellant (gunpowder) to explode

3. Explosion forces bullet out of case

Pintle mount A metal pin holds the gun in a U-shaped bracket, allowing it to tilt up and down. The cylindrical base under the bracket slots into a hole in a tripod, or in a bracket on a vehicle, so the gun can swivel. The whole assembly is known as a pintle mount.

Discover how machine guns fire hundreds of bullets per minute by
visiting www.factsforprojects.com and clicking on the web link.

The Browning M2 is the USA's longest
serving weapon. Troops call it by the
nickname "Ma Deuce."

Some medium and heavy machine guns
are water-cooled so they can fire longer
bursts. Otherwise they may overheat and
"cook-off" or misfire a bullet even though
the trigger is not pulled.

The M3M was developed
mainly for use on US
Navy helicopters, with the
code name GAU-21. It
is also now deployed on
several other vehicles and
craft, from Humvee jeeps
and armored cars to
tanks and spy planes.

Barrel thermal cover Machine
guns get very hot with the
continual explosions inside, and
this may limit the time that they
can fire in short bursts. Various
metal parts such as the barrel
thermal cover, or dissipator,
disperse as much of this heat as
possible to the air around.

Barrel In most guns the
inside of the barrel has spiral
grooves, called rifling, which
make the bullet twist or spin
as it passes along and out.
This causes the bullet to fly
much straighter through
the air due to the gyroscope
effect.

The M3M's
effective range
(where its bullets
can still cause
damage) is over
6,500 ft (almost
2,000 m).

Exchangeable
barrel

.50 Caliber or "50-cal"
means the gun's barrel
has an internal diameter,
or caliber, of half an
inch—0.5 in. The metric
equivalent of 12.7 Caliber
(12.7 mm) doesn't sound
quite so cool.

Muzzle The speed at
which the bullet exits
the end of the barrel,
or muzzle, depends on
the ammunition fired.
In some versions of the
M2 and M3 it can be
more than 2,600 feet per
second (800 meters per
second).

The M3M began
tests in 2001 and
went into full service
in 2004. It is 60 in
(152 cm) long and
weighs 80 lbs (36
kg). The barrel
length is 36 in
(91 cm), of which
31.5 in (80 cm) have
eight rifling grooves.

The F-86 Sabre was one of
many M2-equipped aircraft

✳ Airborne FIREPOWER

More than 100 versions of the Browning
M2 have been fired in over 50 wars
and major conflicts. Some of the most
highly modified were fitted to fighter
and bomber aircraft. The US F-86 Sabre
of the 1950s had six AN/M3 guns,
angled so their armor-piercing bullets
converged at one spot about 980 feet
(300 meters) in front of the plane. Every
sixth round (bullet) was a tracer that
glowed brightly so the pilot could see
and aim the stream of ammunition.

TOMAHAWK AND GRANIT MISSILES

Guided missiles are basically bombs that fly under their own power with some kind of onboard or ground-control guidance. The US's Tomahawk launches from ships or submarines and can strike to within a few feet of targets up to 1,500 miles (2,500 km) away. The larger Russian Granit is also sea-launched but much faster, so its range is limited to less than 625 miles (1,000 km).

Eureka!

The V-1 Flying Bomb of World War II (1939–45), also known as the "buzz-bomb" or "doodlebug" from its droning noise, was in effect an early type of cruise missile. It had two small wings and a tail like an aircraft, with a pulse-jet engine at the tail's top.

What next?

Upgrades to the GPS (satellite navigation) set-up may mean that future missiles can be guided to an accuracy of a few inches.

Munitions Various kinds of "bomblets" and other weapons can be carried in up to 24 pre-loaded canisters.

TOMAHAWK MISSILE

The first Tomahawks entered service in 1983 and they are still on standby today.

Nose cone

Gyro

Skin

Radar and guidance Different types of radar detect the surroundings and steer the missile.

✳ How do CRUISE MISSILES work?

The "cruise" of a guided cruise missile is the phase of its flight under its own power, usually from some kind of jet engine, assisted by small wings or fins that provide lift and control direction. Once launched, cruise missiles head toward their target, but they may take detours—either from their on-board computer guidance, or underground control—to avoid problems such as enemy aircraft anti-missile missiles.

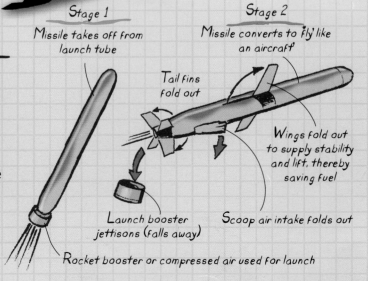

Stage 1
Missile takes off from launch tube

Stage 2
Missile converts to fly like an aircraft

Tail fins fold out

Wings fold out to supply stability and lift, thereby saving fuel

Launch booster jettisons (falls away)

Scoop air intake folds out

Rocket booster or compressed air used for launch

The Tomahawk is about 5.6 m long and weighs almost 1.5 tons. It is subsonic, which means it flies slower than the speed of sound, at about 880 km/h.

Learn all about Tomahawk missiles by visiting www.factsforprojects.com and clicking on the web link.

Older Tomahawks flew at a fixed speed, but improved versions have a throttle to go faster or slower.

At 33ft (10 m) and 7.7 tons (7 metric tons), the Granit is supersonic, reaching over 2,500 mi/h (4,000 km/h).

GRANIT MISSILE

Fin

Engine

Tailplane

Swept wings

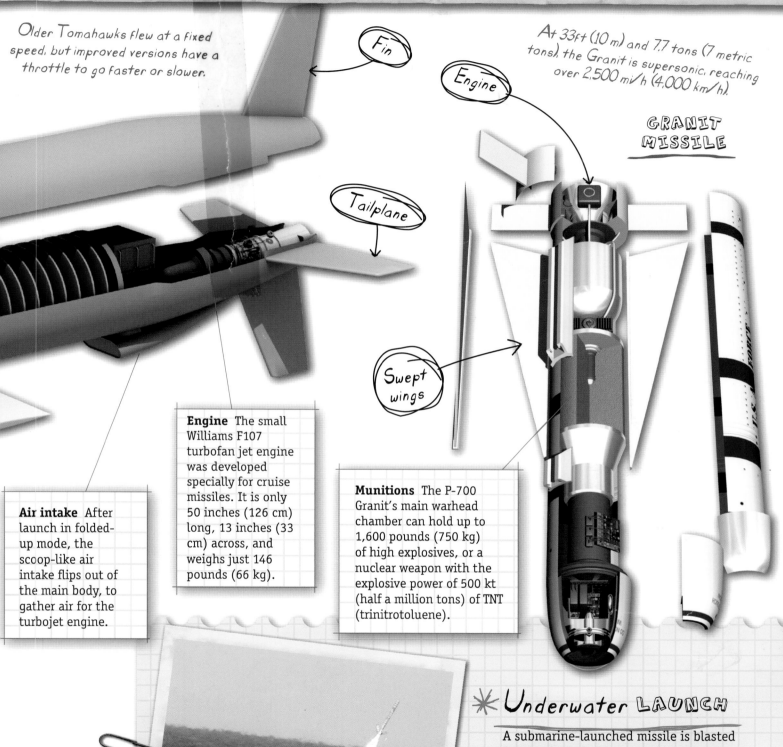

Air intake After launch in folded-up mode, the scoop-like air intake flips out of the main body, to gather air for the turbojet engine.

Engine The small Williams F107 turbofan jet engine was developed specially for cruise missiles. It is only 50 inches (126 cm) long, 13 inches (33 cm) across, and weighs just 146 pounds (66 kg).

Munitions The P-700 Granit's main warhead chamber can hold up to 1,600 pounds (750 kg) of high explosives, or a nuclear weapon with the explosive power of 500 kt (half a million tons) of TNT (trinitrotoluene).

Tomahawks can take off vertically, as well as horizontally, from a submarine's torpedo tubes.

A sub-launched missile booster ignites

✳ Underwater LAUNCH

A submarine-launched missile is blasted from its launch tube, which is also its protective canister during transport, by a sudden burst of extremely high gas or water pressure. After the missile breaks the surface, the solid-fuel rocket booster switches on like a giant firework to gain height and speed as fast as possible. After a preset burn time the booster falls away, then the jet engine fires up and takes over propulsion.

M2 BRADLEY IFV

Part armored car, part troop transport, and part light tank, the M2 Bradley is termed an IFV (Infantry Fighting Vehicle). Its role is to take troops under armored protection to a particular area of the battlefield, support them by providing covering fire and shelter, and repel attacks against itself. Its cousin the M3 is a scout version to survey and gather information.

The Bradley is named after US General Omar Bradley, one of the main commanders in Europe during World War II. It is one of several types of armored troop carriers known as "battle taxis."

Eureka!

Battle-ready soldiers need all their stamina and strength when they get to the front line, rather than having to march there. Early troop carriers were horse-drawn carts and even oxen-pulled wagons.

What next?

Several armies are testing robot-type soldiers. These are not so much human-shaped warriors as very small armored tanks, some no larger than a pet dog—but with tremendous weaponry.

Main gun Most M2 Bradleys are equipped with the M242 Bushmaster chain gun. It has a caliber of 1 inch (2.5 centimeters) and fires shells more than 1.8 miles (3 kilometers).

Engine The Cummins VTA-903T turbo diesel engine is almost 4 gallons (15 liters)—ten times the size of a smallish family car. On full tanks of 175 gallons (660 liters), the M2's range is around 250 miles (400 km).

✳ How do CATERPILLAR TRACKS work?

Also called crawler tracks or endless belt loops, caterpillar tracks have many rectangular plates linked by hinged joints. Long, ridged gear-type teeth on the drive wheel, which is turned by the engine, fit into gaps between the plates and pull the track round and round. The other wheels are toothless and undriven—return rollers that support the track along the top, and road wheels, which take the vehicle's weight along the base.

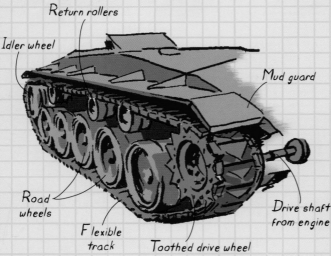

Return rollers

Idler wheel

Mud guard

Road wheels

Flexible track

Toothed drive wheel

Drive shaft from engine

Front armor

Read fact files on the M2 Bradley and other army vehicles by visiting www.factsforprojects.com and clicking on the web link.

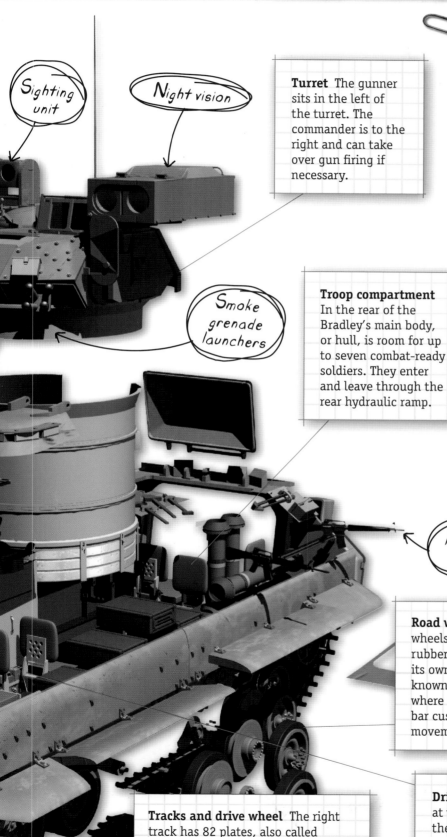

Sighting unit

Night vision

Turret The gunner sits in the left of the turret. The commander is to the right and can take over gun firing if necessary.

An M2 Bradley roars into action

Smoke grenade launchers

Troop compartment In the rear of the Bradley's main body, or hull, is room for up to seven combat-ready soldiers. They enter and leave through the rear hydraulic ramp.

✳ Ready for BATTLE

The M2 Bradley's 600-horsepower diesel engine gives a top speed of 50 miles (65 km) per hour, which is enough to outrun tanks and many other armored vehicles. The tracks can cross most obstacles including ditches and fallen trees. The M6 Linebacker variant carries four Stinger anti-aircraft missiles, while the M7 moves to forward positions and helps direct the firepower of tanks and other attackers.

Machine guns

The basic model of the M2 Bradley is 21 feet (6.5 m) long, 11 feet (3.3 m) wide, and almost 10 feet (3 m) to the top of the turret. When fully equipped and ready for action it weighs about 36 tons (33 metric tons).

Road wheels The six road wheels on each side have rubber tires. Each one has its own suspension system known as a torsion bar, where a springy metal bar cushions the wheel's movements.

Bradleys have been modified into supply carriers, mobile medical centers, and even mobile generators using their own engines or an additional engine-powered electrical generator.

Tracks and drive wheel The right track has 82 plates, also called shoes or links, and the left one has 84. The drive wheel at the front bears 11 teeth or sprockets.

Driver's station This is at the front left, next to the engine. Foot pedals control the engine speed.

M270 ROCKET LAUNCHER

Warhead-carrying rockets and missiles are among the most destructive in battle. The M270 MRLS (Mobile Rocket Launch System) is a speedy tracked launcher based on the M2 Bradley shown on the previous page. Its speciality is "shoot and scoot"—get into position fast, launch its load of up to 12 rockets, and move on before the enemy can pinpoint its position and fire back at it.

Eureka!

Rockets were first used in warfare more than 1,000 years ago in China. They were called "fire arrows" and driven by "black powder," an early form of gunpowder. They caused few casualties, being mainly intended to scare the enemy.

What next?

The M270 will gradually be replaced by the HIMARS (High Mobility Artillery Rocket System). Known as "Six-Pack," it carries half the load but is faster and more accurate.

More than 400 years ago in Korea, large arrows with gunpowder tips were launched by more gunpowder, like big exploding fireworks. More than 150 could be let off from a cart pulled by people (or very scared oxen).

Armor All parts of the M270 are made of armored aluminium-based metal alloys. The sheets are welded for extra strength and to prevent splintering under impact.

The M270 weighs over 26 tons (24 metric tons) but still has a relatively fast top speed of 40 mi/h (64 km/h).

Crew area The standard crew of three consists of the commander on the right, the gunner in the middle and the driver on the left.

Armored roof panel

Blast shutters The front viewing windows are made of specially toughened glass. When under fire, strong shutters fold down over them for extra protection.

LIFT HERE

Drive wheel

✳ Rocket ATTACK

In general warfare, a rocket is self-powered but unguided, while a missile is self-powered and guided. The M270 "family" of ammunition includes the M26, a basic rocket (unguided) that releases grenade-type "submunitions" in mid-air, to shower down and explode on impact. Its range is about 20 miles (32 km). The MGM-140 is a missile (guided) with a range of 200 miles (over 300 km), armed with submunitions or one large warhead.

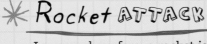

A rocket blasts away from its M270

Watch a video of the M270 firing a volley of rockets by visiting
www.factsforprojects.com and clicking on the web link.

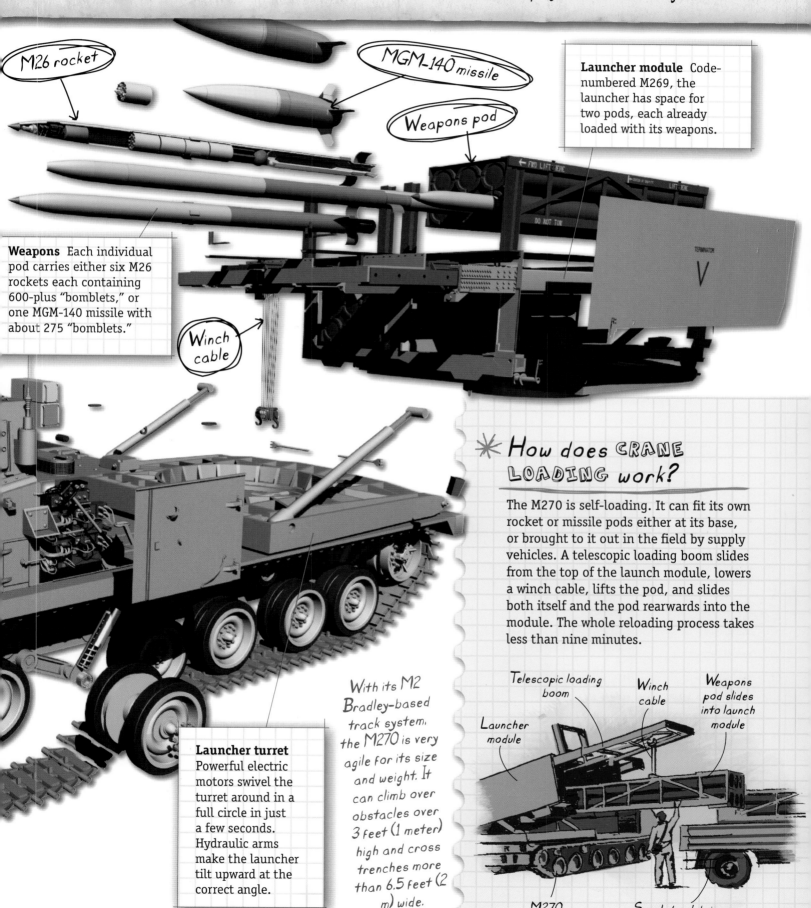

M26 rocket

MGM-140 missile

Weapons pod

Launcher module Code-numbered M269, the launcher has space for two pods, each already loaded with its weapons.

Weapons Each individual pod carries either six M26 rockets each containing 600-plus "bomblets," or one MGM-140 missile with about 275 "bomblets."

Winch cable

Launcher turret Powerful electric motors swivel the turret around in a full circle in just a few seconds. Hydraulic arms make the launcher tilt upward at the correct angle.

With its M2 Bradley-based track system, the M270 is very agile for its size and weight. It can climb over obstacles over 3 feet (1 meter) high and cross trenches more than 6.5 feet (2 m) wide.

✳ How does CRANE LOADING work?

The M270 is self-loading. It can fit its own rocket or missile pods either at its base, or brought to it out in the field by supply vehicles. A telescopic loading boom slides from the top of the launch module, lowers a winch cable, lifts the pod, and slides both itself and the pod rearwards into the module. The whole reloading process takes less than nine minutes.

Telescopic loading boom

Winch cable

Weapons pod slides into launch module

Launcher module

M270

Supply truck brings pods to the M270

15

M1 ABRAMS TANK

One of the world's most powerful mobile weapons, the M1 Abrams MBT (Main Battle Tank) packs a massive punch, and is also protected against almost all kinds of return fire. Its ammunition is stored in a "blowout" compartment so that if it explodes, the force of the blast directs to the outside, rather than within the tank, which would kill the crew at once.

Eureka!

The first tanks went into action in 1916, in the midst of World War I. To keep their function secret from the enemy while they were transported, they were said to be large storage tanks for fresh water. The name stuck.

What next?

Armored protection continually improves. Reactive armor has an explosive layer that detonates when hit, deflecting the incoming force.

Shells

Turret The traverse or swivelling of the turret is carried out by electric motors, but it can be traversed by a hand-cranked handle in an emergency.

Engine The Honeywell AGT 1500C turbine engine is similar to an aircraft jet engine. It produces more than 1,500 horsepower and can run on several types of fuel.

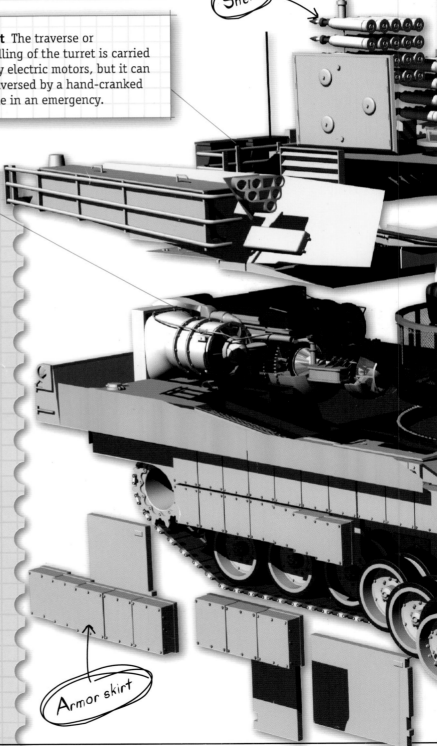

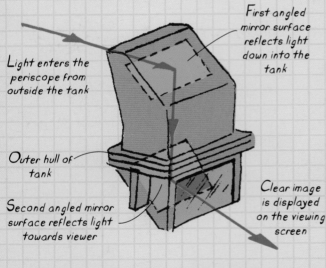

First angled mirror surface reflects light down into the tank

Light enters the periscope from outside the tank

Outer hull of tank

Second angled mirror surface reflects light towards viewer

Clear image is displayed on the viewing screen

✳ How do PERISCOPES work?

When the M1 goes into battle, all hatches are closed. The driver and crew have electronic screens to see outside, but the old technology of the periscope is still essential. The M1 has nine periscopes, also known as vision blocks. Incoming light bounces, or reflects, off two mirror surfaces, each at an angle of about 45°, through a gap in the tank's main outer body covering or hull. The periscope unit can be replaced by a night-vision version using electronics to detect and display infrared (heat) rays.

Armor skirt

>>> MILITARY MACHINES <<<

To discover just how deadly the mighty M1 Abrams is, visit
www.factsforprojects.com and click on the web link.

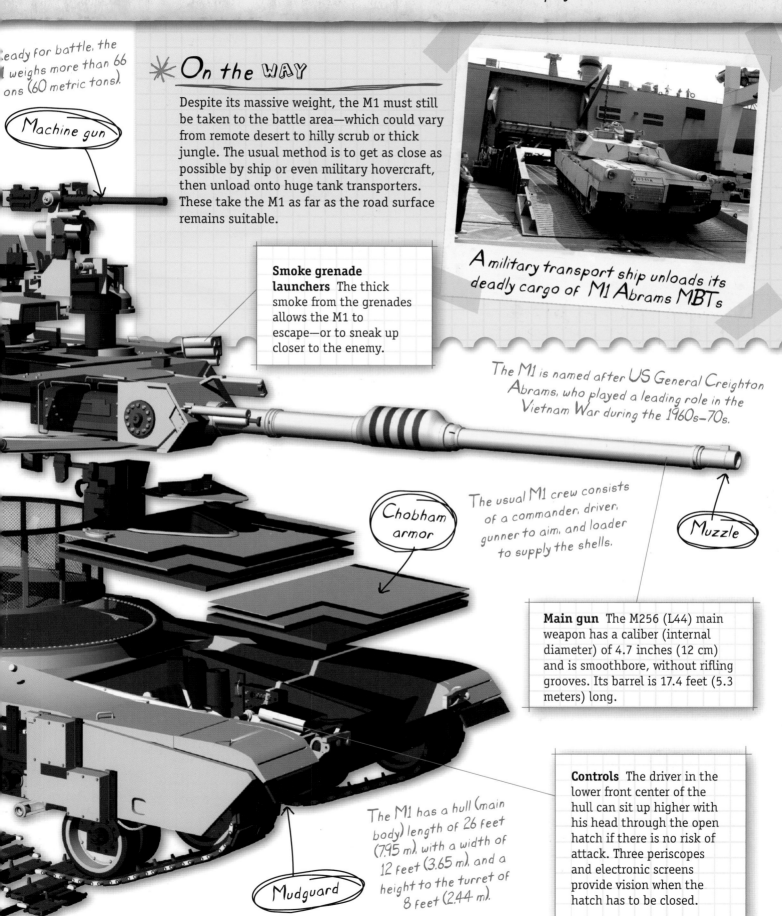

ready for battle, the
weighs more than 66
ons (60 metric tons).

Machine gun

✳ On the WAY

Despite its massive weight, the M1 must still
be taken to the battle area—which could vary
from remote desert to hilly scrub or thick
jungle. The usual method is to get as close as
possible by ship or even military hovercraft,
then unload onto huge tank transporters.
These take the M1 as far as the road surface
remains suitable.

A military transport ship unloads its
deadly cargo of M1 Abrams MBTs

Smoke grenade launchers The thick
smoke from the grenades
allows the M1 to
escape—or to sneak up
closer to the enemy.

The M1 is named after US General Creighton
Abrams, who played a leading role in the
Vietnam War during the 1960s–70s.

The usual M1 crew consists
of a commander, driver,
gunner to aim, and loader
to supply the shells.

Chobham armor

Muzzle

Main gun The M256 (L44) main
weapon has a caliber (internal
diameter) of 4.7 inches (12 cm)
and is smoothbore, without rifling
grooves. Its barrel is 17.4 feet (5.3
meters) long.

The M1 has a hull (main
body) length of 26 feet
(7.95 m), with a width of
12 feet (3.65 m), and a
height to the turret of
8 feet (2.44 m).

Mudguard

Controls The driver in the
lower front center of the
hull can sit up higher with
his head through the open
hatch if there is no risk of
attack. Three periscopes
and electronic screens
provide vision when the
hatch has to be closed.

AH-64 APACHE HELICOPTER

Helicopters are among the most versatile military machines. The Apache is chiefly an attack "helicopter gunship," fitted with guns, rockets, missiles, and other weaponry for use against ground targets. But in a pinch, this fast and agile chopper can carry out surveillance, transport emergency equipment, carry the sick or wounded, and even attend natural disasters to rescue trapped people or bring urgent relief supplies.

Eureka!

A relatively recent addition to military hardware, helicopters first saw action in World War II with the US's Sikorsky X/R-4. This was also the first helicopter to enter mass production, with 130 made.

What next?

Tiny helicopters no larger than a dragonfly are becoming "spies in the skies." Remote controlled, their cameras take pictures to send by radio or deliver back to base.

Radar dome

Canopy The canopy windows, made of specially strengthened glass composite, allow a very wide view, including above.

An Apache banks into a tight turn

✳ High speed ACTION

The Apache has a top speed of almost 190 miles (300 km) per hour and climbs at a stomach-churning rate of 39 feet (12 meters) per second. It is also one of the few helicopters that can "loop the loop," flying up and right over onto its back before diving to become horizontal again in a complete circle. This puts huge strain on the craft, especially its rotor blades. However this happens mainly as a result of the helicopter's moving energy or momentum. It cannot fly upside down in a sustained way, as some aerobatic planes can.

Sights

The Apache weighs 8.8 tons (8 metric tons). It can carry a weapons load of an extra 2-plus tons.

Gun The M230 chain gun has a caliber of 1.2 inches (30 mm). It can be "locked" to point the way the helicopter faces, or aimed independently via the co-pilot-gunner's head-up helmet display.

Crew The co-pilot-gunner sits in the lower front position, with the main sights and weapon controls, with the pilot above and behind.

To become an expert on the AH-64 Apache,
visit www.factsforprojects.com and click on the web link.

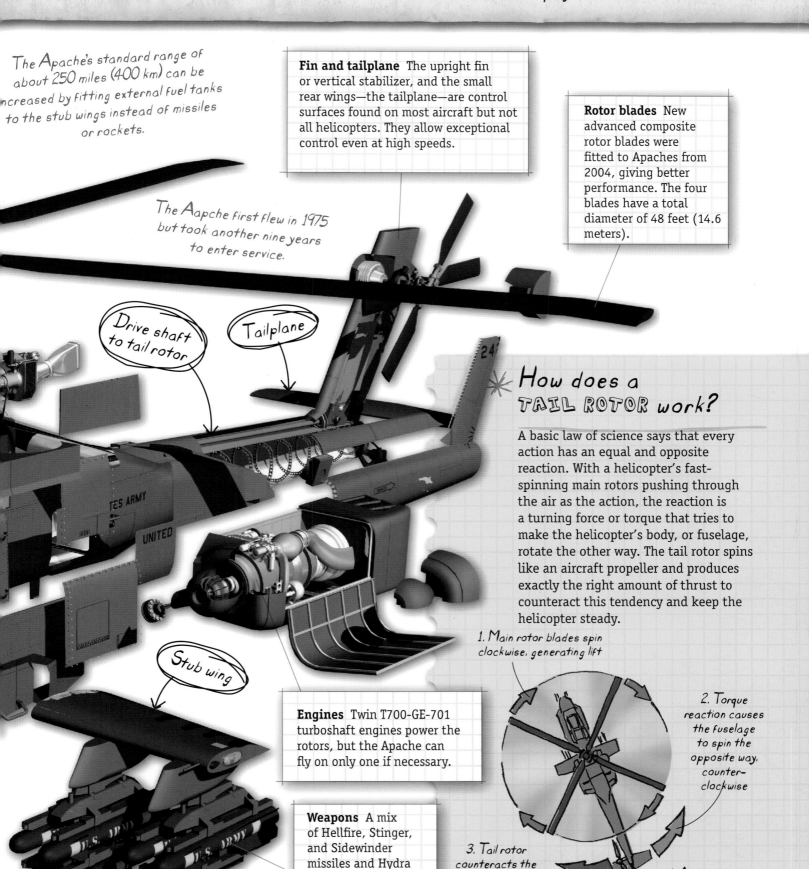

The Apache's standard range of about 250 miles (400 km) can be increased by fitting external fuel tanks to the stub wings instead of missiles or rockets.

The Aapche first flew in 1975 but took another nine years to enter service.

Fin and tailplane The upright fin or vertical stabilizer, and the small rear wings—the tailplane—are control surfaces found on most aircraft but not all helicopters. They allow exceptional control even at high speeds.

Rotor blades New advanced composite rotor blades were fitted to Apaches from 2004, giving better performance. The four blades have a total diameter of 48 feet (14.6 meters).

Drive shaft to tail rotor

Tailplane

Stub wing

Engines Twin T700-GE-701 turboshaft engines power the rotors, but the Apache can fly on only one if necessary.

Weapons A mix of Hellfire, Stinger, and Sidewinder missiles and Hydra rockets can be clipped to the stub-wings.

How does a TAIL ROTOR work?

A basic law of science says that every action has an equal and opposite reaction. With a helicopter's fast-spinning main rotors pushing through the air as the action, the reaction is a turning force or torque that tries to make the helicopter's body, or fuselage, rotate the other way. The tail rotor spins like an aircraft propeller and produces exactly the right amount of thrust to counteract this tendency and keep the helicopter steady.

1. Main rotor blades spin clockwise, generating lift

2. Torque reaction causes the fuselage to spin the opposite way, counter-clockwise

3. Tail rotor counteracts the torque reaction so the fuselage remains stable

4. Adjusting tail rotor speed allows helicopter to turn while hovering

A-10 THUNDERBOLT

Some of the affectionate names given to the A-10 by its pilots and engineers are "Wart-hog," "Hog," and "Tankbuster." This relatively slow but exceptionally tough, well-armored aircraft offers CAS (Close Air Support). It is called in to attack enemy troops and ground vehicles near its own forces. Despite its limited speed, the A-10 is extremely agile, twisting and turning to evade anti-aircraft guns and missiles.

Eureka!

The A-10's GAU/8A Avenger Gatling gun, or rotary canon, is one of the most powerful guns ever fitted to an aircraft. The original version was invented in 1861 by Richard Gatling (1818–1903) at the start of the American Civil War (1861–65).

What next?

The "good old A-10" is near the end of its life. In the 2020s it should be replaced by the F-35 Lightning, which first flew in 2006.

Ejector sea

The A-10 has a very long service history. It first flew in 1972 and it may still be around 50 years later.

Dash

Main gun

Nose cone

✳ Nice 'n' SIMPLE

The A-10 is not fitted with the latest complicated machinery and features—and that is part of its success. It is designed to still be able to fly with just one engine, as well as with one tailplane plus part of a main wing missing. The jet engines are mounted on stub pylons rather than being built into the wings or fuselage. This means their covers can be removed for fast, easy servicing and repair.

Armored bathtub
The pilot sits in a protective container made of lightweight but super-tough titanium metal known as the "bathtub."

The exposed engines of the A-10 are easy to work on or replace

Bearing

Rotary cannon The massive cannon fires a standard mix of armor-piercing and high-explosive shells. They leave at a speed of almost 3,280 feet (1,000 meters) per second.

Ammunition drum The drum holds more than 1,150 rounds. Each round, or shell, is almost 1 foot (30 cm) long.

The recoil or "kick" of the rotary canno at the instant each shell fires is about the same as the aircraft's two engines

Find out all about the A-10's huge cannon by visiting
www.factsforprojects.com and clicking on the web link.

Only one two-seater A-10 was built, as a test aircraft for night-time and bad-weather work.

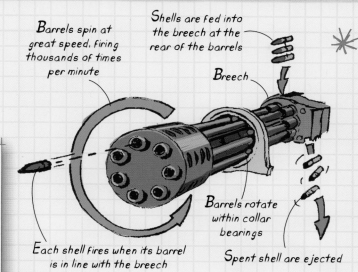

Barrels spin at great speed, firing thousands of times per minute

Shells are fed into the breech at the rear of the barrels

Breech

Each shell fires when its barrel is in line with the breech

Barrels rotate within collar bearings

Spent shell are ejected

✳ How does a ROTARY CANNON work?

A Gatling-type gun reloads itself rapidly but not under its own power. It must be cranked, or turned, by an external force such as a person, an electric motor, or a hydraulic (pressurized liquid) or pneumatic (pressurized air) mechanism. The A-10's Avenger gun has seven barrels, measures nearly 20 feet (6 m) in length and fires 65 times per second. With its full load of ammunition, it weighs almost 2.2 tons (2 metric tons).

Engines The twin TF34-GE-100 turbofan jet engines give the A-10 a top speed of 435 miles (700 km) per hour.

Twin fins The two fins, each with a rudder, allow the A-10 to turn very sharply, even at low speeds when the air flowing past has little pushing force on the rudder.

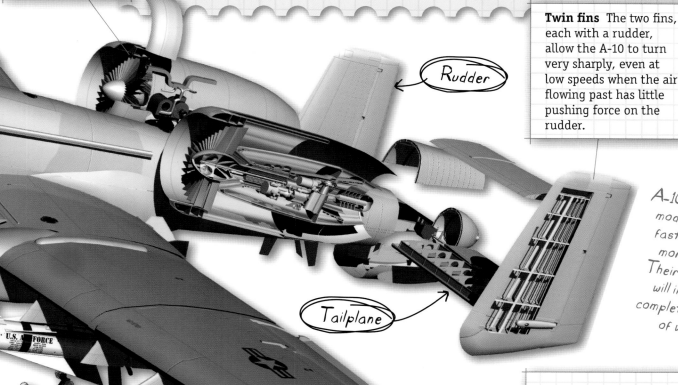

Rudder

A-10s are being modified to go faster and be more lethal. Their upgrades will include a complete new set of wings!

Tailplane

High-lift wing The large wing area, along with its highly curved top surface, mean the wing creates plenty of lift force even at low speeds.

Weapons The A-10 is armed with Sidewinder and Maverick missiles plus a range of bombs attached to its 11 clip-on "hardpoints."

name "Thunderbolt" honors the P-47 Thunderbolt propeller-driven fighter of World War II.

V-22 OSPREY TILTROTOR

For many years, inventors have tried to build an aircraft with the benefits of both a standard fixed-wing plane and a helicopter. The V-22 Osprey is one of the more successful ones. The engine-and-rotor assembly at each wingtip swivels so the rotor faces up for vertical take-off or landing, then forward for normal flight. Ospreys can carry cargoes of more than 6.6 tons (6 metric tons) for hundreds of miles which no helicopter can match.

Eureka!

Tiltrotor designs were sketched in the 1930s but never built. A German prototype, the FA-269 "Heliplane," was partly built in the 1940s but never flew. The Osprey was developed from the Bell XV-15, which first flew in 1977.

What next?

The HTR (Hybrid Tandem Rotor) is a planned combination of tiltrotor and helicopter—a tiltwing. Its whole wing swivels by 25°, but only on the drawing board. Construction has not yet started.

The Osprey's wingspan is 46 feet (14 m) but the rotors increase its width to almost 85 feet (26 m).

Stub wing The wing must be much stiffer and stronger than in an ordinary plane, to carry the weight of the engine and rotor and withstand their pulling force.

Nacelle

✳ How do TILTROTORS work?

Helicopters have amazing flight abilities, especially hovering. However they lack wings to give them lift as they move forward. Without this lift, they have to use much more fuel to stay airborne. The V-22 Osprey solves this by having aircraft wings, and also rotors that can be tilted and controlled like those of a helicopter. However, in helicopter mode the wing gets in the way of the air downflow from the rotors, reducing their effectiveness.

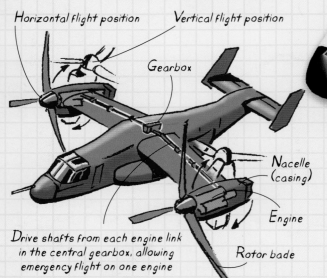

Horizontal flight position   Vertical flight position

Gearbox

Nacelle (casing)

Engine

Rotor bade

Drive shafts from each engine link in the central gearbox, allowing emergency flight on one engine

Crew and controls The flight deck crew of two are the usual pilot in the left seat and co-pilot in the right. The controls are more unusual, including rotor tilt angle, but computers do much of the work automatically.

In the main fuselage there is room for up to 32 troops. A hook beneath allows the Osprey to airlift vehicles.

Check out the V-22 Osprey program's official website by visiting www.factsforprojects.com and clicking on the web link.

Rotor hub This complicated device, as found on a helicopter, adjusts the angle of the rotor blades while hovering, depending on the amount of lift needed.

The early development of the Osprey was marred by several crashes and the loss of more than 30 lives, needing much redesign. By the time the aircraft got to test flight stage, its cost had increased ten times.

Rotor blades A combination of aircraft propeller and helicopter rotor, the rotor blades are called proprotors. Each has three blades and an overall diameter of 38 feet (11.6 meters). Because of the two jobs they do, the blades are not especially efficient at either.

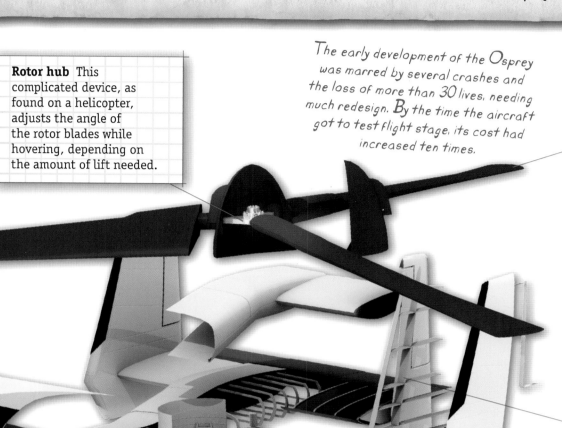

Airframe The very light structural framework and cladding mean the Osprey has a range of some 995 miles (1,600 km), increasing to more than 2,500 miles (4,000 km) with mid-air refueling.

✳ Pack-up PLANE

If there's no rush, it's far more fuel-efficient to take a V-22 long distances by ship. To stow away the craft, the whole wing and engine-rotor assemblies swivel around by 90° so the wing is in line with the fuselage. At each wing tip, two of the rotor blades can also fold to line up with the third.

Engine Each Rolls Royce TT406/AE 1107 turboshaft engine puts out more than 6,000 horsepower. The pumps for the fuel and other flowing liquids must work whether the engine is vertical or horizontal, counteracting the effects of gravity.

The Osprey folds up for easy storage

F-22 RAPTOR

Some of the meanings for the term "raptor" are "thief," "hunter" and "raider"—and the F-22 is all of these. A very rapid and adaptable aircraft, it uses stealth design to sneak into enemy air space, gather survey information, listen to enemy radio messages, jam radar, carry out a missile strike, and blast away again using its afterburners before even being detected. All for just $143 million, plus the cost of the pilot and fuel.

Eureka!

Stealth technology began in the late 1950s with aircraft designed not to reflect, or bounce back, the radio waves of radar. It has since spread to lessen a craft's noise levels, heat production, radio messages and even visible shape.

What next?

In active camouflage or active stealth, an object "bends" light from the background around itself and on toward the observer, and so becomes invisible.

Lt Col. Gary Jeffrey

Canopy

Fly-by-wire Like most modern jets, the side-stick control is linked to computers that send instructions as electrical signals along wires, to move parts such as the rudder and elevators. This replaces the old metal cables that physically pulled these parts.

Radar A special type of stealth radar known as AESA (Active Electronically Scanned Array) makes the F-22 difficult to detect.

Probe

Engine air intake

✳ BURST of speed

Afterburning, also called reheat, greatly increases a jet engine's pushing force, or thrust. However this uses huge amounts of extra fuel. It also produces a lot of extra heat that the engine and surrounding structure cannot withstand for long. So it's used only when necessary. One situation is during take-off, to reach lift-off speed as quickly as possible. In combat, afterburners allow the craft to accelerate away from danger.

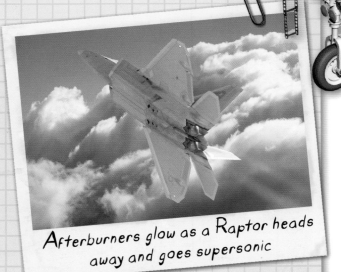

Afterburners glow as a Raptor heads away and goes supersonic

The Raptor first took to the air in 1997. However some of its roles can be partly carried out by other aircraft, such as the F-35. The cost of each F-22 also rose to $143 million. By the late 2000s, its future was uncertain.

For news about the Raptor and its actions around the world, visit www.factsforprojects.com and click on the web link.

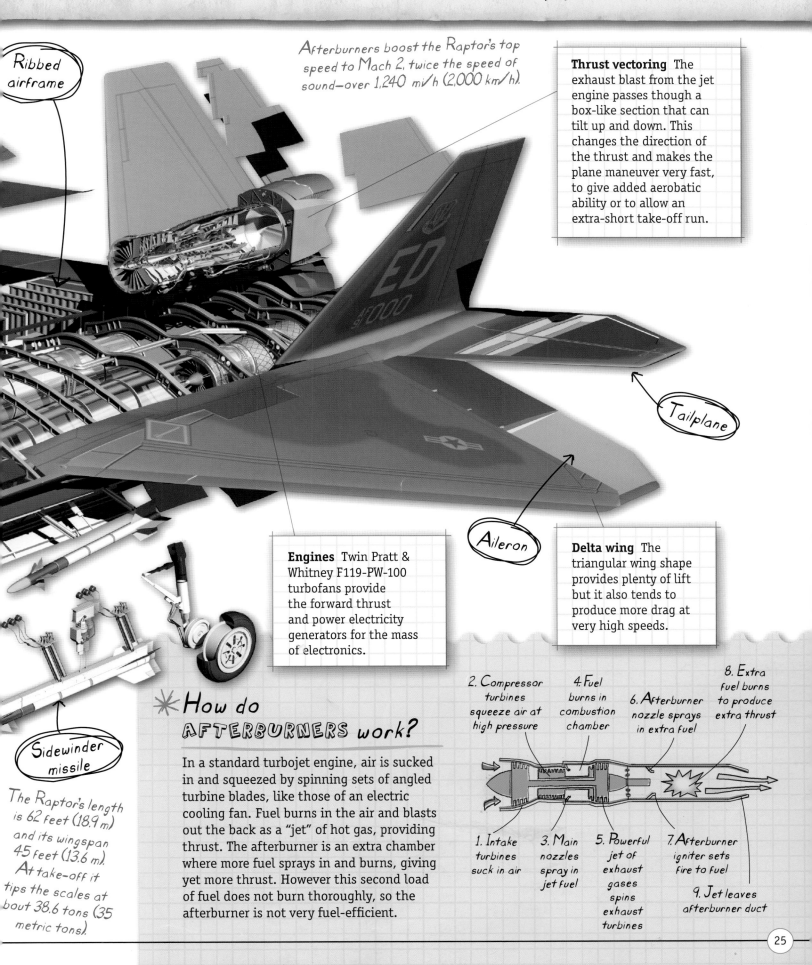

Ribbed airframe

Afterburners boost the Raptor's top speed to Mach 2, twice the speed of sound—over 1,240 mi/h (2,000 km/h).

Thrust vectoring The exhaust blast from the jet engine passes though a box-like section that can tilt up and down. This changes the direction of the thrust and makes the plane maneuver very fast, to give added aerobatic ability or to allow an extra-short take-off run.

Tailplane

Aileron

Engines Twin Pratt & Whitney F119-PW-100 turbofans provide the forward thrust and power electricity generators for the mass of electronics.

Delta wing The triangular wing shape provides plenty of lift but it also tends to produce more drag at very high speeds.

Sidewinder missile

The Raptor's length is 62 feet (18.9 m) and its wingspan 45 feet (13.6 m). At take-off it tips the scales at bout 38.6 tons (35 metric tons).

✳ How do AFTERBURNERS work?

In a standard turbojet engine, air is sucked in and squeezed by spinning sets of angled turbine blades, like those of an electric cooling fan. Fuel burns in the air and blasts out the back as a "jet" of hot gas, providing thrust. The afterburner is an extra chamber where more fuel sprays in and burns, giving yet more thrust. However this second load of fuel does not burn thoroughly, so the afterburner is not very fuel-efficient.

2. Compressor turbines squeeze air at high pressure

4. Fuel burns in combustion chamber

6. Afterburner nozzle sprays in extra fuel

8. Extra fuel burns to produce extra thrust

1. Intake turbines suck in air

3. Main nozzles spray in jet fuel

5. Powerful jet of exhaust gases spins exhaust turbines

7. Afterburner igniter sets fire to fuel

9. Jet leaves afterburner duct

E-3 SENTRY AWACS

Modern military conflicts are partly about hardware, such as tanks, battleships, planes, and missiles—and partly about electronics. Radio and microwave communications, radar and satellite systems are vital tools. Airborne Warning And Control Systems (AWACS) such as the Sentry keep watch for suspicious activity and also command and coordinate their own forces.

Eureka!

In radar (see opposite), radio waves can detect an object's direction, distance and perhaps its identity—aircraft, ship, missile, or rocket. Radar was explored by several inventors in the 1930s. The first working systems were devised by an Allied team led by Robert Watson-Watt in World War II.

What next?

Satellite radar and madar (based on microwaves rather than radio) may one day track almost every single military vehicle and craft.

The Sentry is 150 feet (46 m) long, 41 feet (12.5 m) high to the fin tip, and has a wingspan of almost 145 feet (44 m).

The Sentry can stay in the air for more than eight hours without refueling. Its best cruising speed, with greatest fuel economy, is about 350mi/h (570 km/h).

Radar operators study their screens during an AWACS trip

Navigation The Sentry has six main navigation systems including ground-mapping radar, GPS—and the old-fashioned magnetic compass in case of emergency.

Computer stations Up to 19 mission crew operate monitor stations, each with screens and displays. These show the multitude of invisible radio, radar and microwave signals always present in the atmosphere.

✳ Eye in the Sky

The E-3 Sentry is based on one of the first and longest-serving passenger jets, the Boeing 707. Sentries patrol the skies around the clock, monitoring air activity with their amazingly powerful radar. They also "listen in" to radio and microwave signals used by other nations. All of these signals are screened by computers programmed to alert the operators to anything unusual.

Airframe formers

Flight deck

Flight radar

NATO ⊕ OTAN

Onboard electronics Vast amounts of electronic and computing equipment mean that a significant amount of jet engine power is used to generate electricity.

Main landing gear

To read a factsheet and watch a video on the E-3 Sentry visit
www.factsforprojects.com and click on the web link.

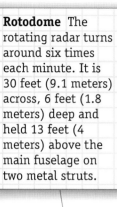

Rotodome The rotating radar turns around six times each minute. It is 30 feet (9.1 meters) across, 6 feet (1.8 meters) deep and held 13 feet (4 meters) above the main fuselage on two metal struts.

In 1995, a Sentry crashed in Alaska. The cause was traced to a flock of Canada geese that were sucked into two of the engines.

"Lookdown" radar The immensely powerful million-watt pulse Doppler radar detects low-flying aircraft, ships and similar objects more than 186 miles (300 kilometers) away. Its range is even farther for high-flying planes—and spacecraft.

✴ How does PULSE DOPPLER RADAR work?

Radar bounces radio signals off an object and detects the echoes, to work out its direction and distance. With short signals or pulses, if the object moves toward them, it gets slightly nearer for each pulse, so the reflected pulses are closer together. Similarly if the object moves away, the pulses are more spaced out. This Doppler effect allows fast-pulse radar to measure an object's speed.

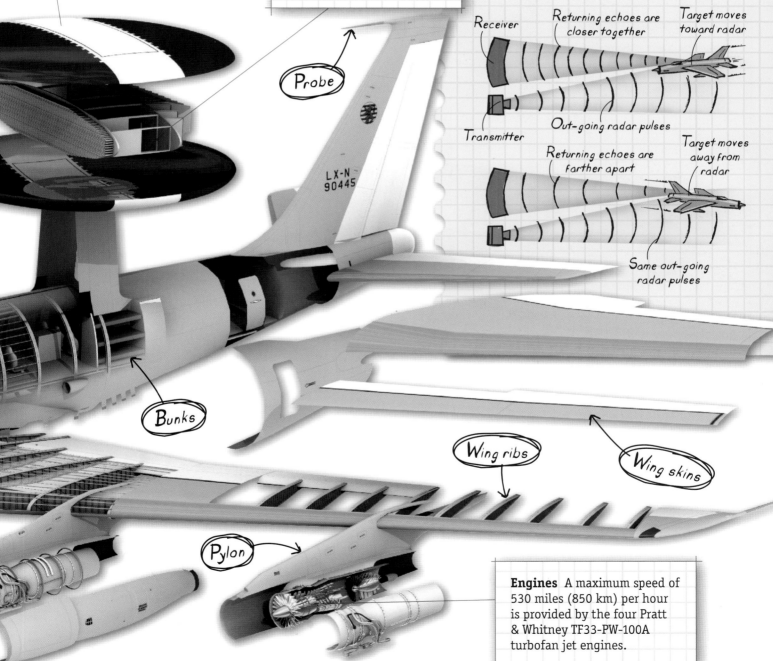

Receiver

Returning echoes are closer together

Target moves toward radar

Transmitter

Out-going radar pulses

Returning echoes are farther apart

Target moves away from radar

Same out-going radar pulses

Probe

LX-N 90445

Bunks

Wing ribs

Wing skins

Pylon

Engines A maximum speed of 530 miles (850 km) per hour is provided by the four Pratt & Whitney TF33-PW-100A turbofan jet engines.

B-52 STRATOFORTRESS

Stalwart of the US Air Force for more than 50 years, the B-52 heavy bomber was designed for the Cold War era of the 1950s–70s. At this time the superpowers of the USA and former USSR (Russia and its allies) paraded their nuclear weapons with unspoken but clear threats. The B-52 not only carries a massive weapons load, it can deliver to targets almost 5,000 miles (8,050 km) away and return with ease.

Eureka!

The first purpose-designed large aircraft to drop bombs saw action in World War I. For their heavy loads they needed multiple engines, from two to four or more, compared to the small, light, single-engined fighters.

What next?

Solar-powered, remote-controlled aircraft can stay up for days, and may be used to drop small, powerful, lightweight explosives.

Crew The basic crew of five are headed by the captain-pilot, along with the co-pilot, navigator, bombardier or "target acquisition officer," and an electronics warfare specialist.

Weapons bay The total weight of weapons can be up to 35 tons (32 metric tons), depending on the fuel load. The bombs, rockets, missiles and mines are mixed and matched according to the mission.

Massive wing The enormous wings give plenty of lift so that less fuel is needed to keep the B-52 airborne.

The B-52's nose shape lent its name to a hairstyle popular in the 1960s. Also known as the "beehive" it has been revived by performers such as Amy Winehouse.

Twin engine pods

A B-52 takes a high-altitude "drink"

✳ Topping-UP up TOP

Mid-air (aerial or in-flight) refueling was first achieved in 1923 in the USA, between two DH-4B aircraft. By 1930 records of more than 500 hours in the air were being set by teams of pilots and flight crew. For a truly long-distance mission the refueling plane, or tanker, is itself refueled by a second tanker, and this can extend to a third tanker, in a relay system with carefully timed flights to meeting points or rendezvous locations.

The B-52 is not especially fast compared to modern aircraft, with a top speed of 620 mi/h (1,000 km/h). Its advantages are its long range and great destructive load.

Read about the history as well as the current operations for the B-52 by visiting www.factsforprojects.com and clicking on the web link.

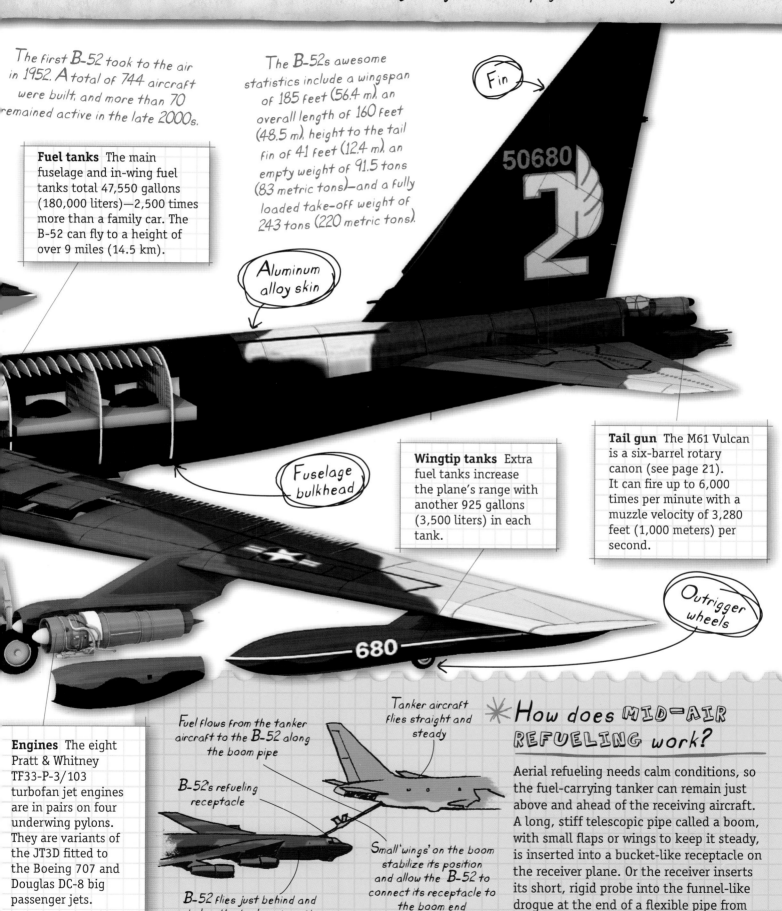

The first B-52 took to the air in 1952. A total of 744 aircraft were built, and more than 70 remained active in the late 2000s.

The B-52s awesome statistics include a wingspan of 185 feet (56.4 m), an overall length of 160 feet (48.5 m), height to the tail fin of 41 feet (12.4 m), an empty weight of 91.5 tons (83 metric tons)—and a fully loaded take-off weight of 243 tons (220 metric tons).

Fin

50680

2

Fuel tanks The main fuselage and in-wing fuel tanks total 47,550 gallons (180,000 liters)—2,500 times more than a family car. The B-52 can fly to a height of over 9 miles (14.5 km).

Aluminum alloy skin

Fuselage bulkhead

Wingtip tanks Extra fuel tanks increase the plane's range with another 925 gallons (3,500 liters) in each tank.

Tail gun The M61 Vulcan is a six-barrel rotary canon (see page 21). It can fire up to 6,000 times per minute with a muzzle velocity of 3,280 feet (1,000 meters) per second.

Outrigger wheels

680

Engines The eight Pratt & Whitney TF33-P-3/103 turbofan jet engines are in pairs on four underwing pylons. They are variants of the JT3D fitted to the Boeing 707 and Douglas DC-8 big passenger jets.

Fuel flows from the tanker aircraft to the B-52 along the boom pipe

Tanker aircraft flies straight and steady

B-52's refueling receptacle

Small 'wings' on the boom stabilize its position and allow the B-52 to connect its receptacle to the boom end

B-52 flies just behind and below the tanker aircraft

✳ How does MID-AIR REFUELING work?

Aerial refueling needs calm conditions, so the fuel-carrying tanker can remain just above and ahead of the receiving aircraft. A long, stiff telescopic pipe called a boom, with small flaps or wings to keep it steady, is inserted into a bucket-like receptacle on the receiver plane. Or the receiver inserts its short, rigid probe into the funnel-like drogue at the end of a flexible pipe from the tanker.

AVENGER MINEHUNTER

Naval mines are floating or seabed bombs that explode when another craft comes near or touches them. They are cheap to make and release or "lay," and remain a threat for months or even years to ships and boats in the area. Minehunters like the Avenger patrol an area, detect mines using sonar equipment, and destroy them.

All 14 of the the US's Avenger-class vessels are named after the first one built. They are relatively small, at 223 feet (68 m) long and with a beam (maximum width) of 39 feet (12 m). Ready-for-action weight is 1,433 tons (1,300 metric tons).

Eureka!

The first sonar or echo-sounding systems came into use in 1914. They followed the terrible tragedy of the sinking of the *Titanic* in 1912, when sonar could have warned the liner about the iceberg.

What next?

Dolphins and some whales use a natural form of sonar to find their way. Dolphin military "recruits" can be trained to detect mines and may carry out this role in the future.

Mast Apart from radar to detect aircraft, other ships and perhaps mines, the mast carries several types of antenna (aerials) for radio and microwave communications.

Hull The main hull structure is wood—chiefly oak, fir and cypress—covered with a skin of glass-fiber plastic. Lack of metal means a low "magnetic signature," which is much less likely to set off magnetic mines.

Sonar pods are readied for underwater action

✳ Seeing with SOUND

Mines and other objects in the water can be detected by active sonar on the minehunter vessel, usually mounted on its hull. The returning echoes are analyzed and displayed by computers on screens. Several transmitters and receivers spaced apart on the hull, known as a sonar array, give a more accurate view, because echoes returning from a particular direction reach the nearest receivers first. Or sonar equipment can be towed on long lines behind a vessel, to lessen interference from its own engines and other noises.

Wooden ribs

Deck

Bow

Find out about the incredible history of military dolphins by visiting www.factsforprojects.com and clicking on the web link.

USS Avenger began construction in 1983, was launched in 1985, and received its commission into service in 1987.

Funnel

Display screens

✴ How does SONAR work?

Sonar (Sound Navigation And Ranging) is the sound version of radar. In active sonar, a vessel sends out ping-like pulses of sound, which travel well and far through water. The vessel's underwater microphones, called hydrophones, detect any returning echoes and a computer works out the size, distance, and direction of the object. In passive sonar the vessel simply "listens" for noises in the water—from whales to enemy subs.

Active Sonar

Hydrophone receiver "hears" the echoes

Transmitter sends out sound pulses

Sound pulses hit objects and bounce back or reflect as echoes

Passive Sonar

Hydrophone on a long line is clear of noise produced by its own ship's engines

Objects such as submarine engines produce noise and vibrations that pass easily through water

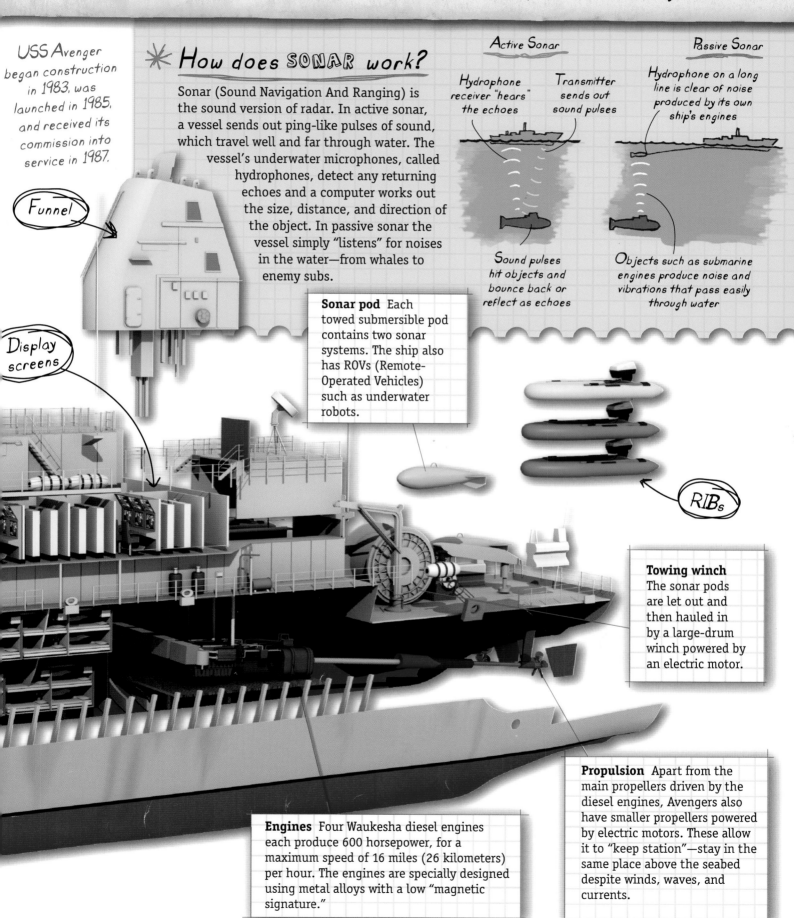

Sonar pod Each towed submersible pod contains two sonar systems. The ship also has ROVs (Remote-Operated Vehicles) such as underwater robots.

RIBs

Towing winch The sonar pods are let out and then hauled in by a large-drum winch powered by an electric motor.

Engines Four Waukesha diesel engines each produce 600 horsepower, for a maximum speed of 16 miles (26 kilometers) per hour. The engines are specially designed using metal alloys with a low "magnetic signature."

Propulsion Apart from the main propellers driven by the diesel engines, Avengers also have smaller propellers powered by electric motors. These allow it to "keep station"—stay in the same place above the seabed despite winds, waves, and currents.

TYPE 45 DESTROYER

Destroyers are medium-sized, fast, maneuverable, long-distance warships that often escort larger vessels such as aircraft carriers or transporters. They protect these bigger ships against attack from enemy surface vessels, submarines, aircraft, missiles, and other threats. The UK's Type 45 Class destroyers are also known as D Class after the first of their kind, HMS *Daring*, launched in 2006.

Eureka!

All warships were wooden until the arrival of ironclads, the first being France's *La Gloire* (1859), the wooden hull of which was covered in metal sheets. Fully metal hulls came about ten years later.

What next?

The same features used in stealth aircraft such as the F-22 Raptor are being extended to ships, with shapes and materials that lessen reflection of radar's radio waves.

Radar

Engines Two Rolls Royce WR21 gas turbines provide turning power, not for propellers, but for electricity generators.

Drive shafts The engines produce electricity for two massive electric motors that turn drive shafts with propellers at the ends.

HMS Daring was launched in 2006 and went into sea trials and entered service in 2009.

HELI-PAD

Rear deck

Bulkeads

Outer hull

Superstructure The "clean" outline has smooth panels with anti-radar coatings, few right angles and sharp edges, and other stealth features. These reduce the chances of being detected by radar.

✳ How do GAS TURBINES work?

The gas turbine is similar to the turbojet engine shown on page 25. However the propelling force is not the exhaust jet of hot gases. It is the inner shaft in the middle of the engine, which is turned by the compressor turbines. A series of gears slows the rotation speed to spin the ship's propellers with great power.

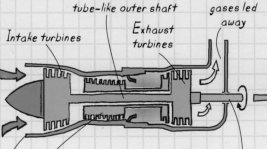

Exhaust turbines connect to intake turbines via tube-like outer shaft

Intake turbines

Exhaust turbines

Exhaust gases led away

Air intake

Compressor turbines

Compressor turbines spin on inner shaft that is connected to gearbox and propellers

HMS Daring honors the destroyer of the same name launched in 1949—and five previous "Darings" in the UK's Royal Navy.

Take a guided tour of a state-of-the-art Type 45 Destroyer
by visiting www.factsforprojects.com and clicking on the web link.

Radar

Funnel The funnel has several cooling systems for the hot exhausts from the gas turbines. This lowers the ship's "infrared signature," where others can locate it using heat-sensitive or infrared-detecting equipment.

PLAN VIEW

Type 45s are sizeable craft, with a length of 500 feet (152 m), a width of 70 feet (21.2 m), and a total weight of 8,100 tons (7,350 metric tons) when fully equipped for a lengthy voyage.

Bridge

Armament The front gun is a 4.5 Mark 8 (caliber 4.5 inches or 113 millimeters). However the main weapons are a variety of missiles.

T45

Type 45s are built at three separate shipyards, two in Glasgow and one in Portsmouth, with final fitting in Glasgow.

Anchor

Lifeboats Even the lifeboats are hidden behind quick-release panels, so the radio waves of radar cannot bounce off them in a characteristic pattern.

The standard Type 45 has a crew of 190, with room for an extra 40 people if necessary.

✳ SHIP SHAPE

Most ships and boats are built on dry land. Some are constructed on the shore, then launched by sliding down a ramp or slipway into the water. Another method is the dry dock, which is an area that can be closed off with watertight gates and then emptied by pumps or drains. The vessel is launched by flooding the dock—allowing water back in through gaps or pipes—and then opening the gates. Ships and boats also come into dry dock for repairs and refits.

The latest destroyer nears completion on the slipway

TYPE 212 SUBMARINE

Nuclear reactors do not need air to burn their fuel, unlike jet, diesel, or petrol engines. So nuclear-powered submarines can stay underwater and on the move for weeks, even months. The German Type 212 submarine is non-nuclear and has a diesel engine. However it can also travel submerged for weeks using a second source of energy—electricity made by its hydrogen-oxygen-powered fuel cells.

Type 212s have been tested to depths of 3,000 feet (700 m), which is deeper than many other submarines.

Eureka!

A sub's chief weapons include its torpedoes—underwater missiles like self-propelled bombs. The first torpedoes were tested in the 1860s. Some were powered by jets of compressed air from a high-pressure cylinder. Others had wind-up clockwork motors!

The Type 212 can stay under the surface for three weeks. This time is extended to 12 weeks if it comes near the surface to 'snorkel' air.

What next?

Sneaky submarines can come and go undetected, apart from by sonar, as shown previously. Strung-out automatic sonar arrays may form "listening walls" around ports, naval centers, and other important bases.

Sail The fin or sail runs with gentle, smooth curves into the main hull. This reduces the sub's "sonar signature."

Hatches A sub's "doors," or hatches, are normally kept closed underwater. To allow divers out, the diver goes through the inner hatch door into the hatch compartment. The inner door is sealed, the compartment is flooded, and the outer door opens.

Open hatch

✳ How do HYDROPLANES work?

Submarine controls are similar to those of an aircraft. The vessel turns left or right, called yaw, by angling its vertical control surfaces—rudders—to the opposite side. They press against the water flowing past and are pushed sideways, as in a surface ship. A sub tilts up or down, known as pitch, using horizontal control surfaces known as hydroplanes or diving planes.

Moving the rudder to the left (or right) pushes the rear of the sub to the right (or left) for steering

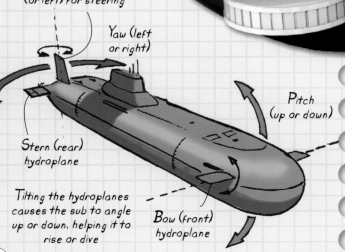

Yaw (left or right)

Stern (rear) hydroplane

Tilting the hydroplanes causes the sub to angle up or down, helping it to rise or dive

Pitch (up or down)

Bow (front) hydroplane

Torpedoes There are six torpedo tubes in two groups of three, and 12 torpedoes ready to fire. The advanced DM2A4 torpedo is 22 feet (6.6 meters) long and driven by an electric motor and batteries, with a range of 31 miles (50 km).

Bathroom

Before refueling, t Type 212 can sa for a distance of more than 416 mile (670 km). Its maximu submerged speed is 23 mi/h (37 km/h).

To watch a video about the Type 212, visit
www.factsforprojects.com and click on the web link.

Periscope The sub's slide-up periscope tube allows a view above the surface of the water.

✳ Prepare to DIVE!

The control center of a ship is usually known as the bridge. On surface vessels it has wide windows for a panoramic view. Underwater, submarines "see" what's around using sonar, as shown previously, as well as other sensors, such as magnetic detectors for metal parts of nearby ships. If the sub is just below the surface, the crew can raise or hoist the extendable periscope so that its upper end is just above the water. This gives an all-round or 360° view.

On the submarine's bridge, the captain keeps an eye on his vessel and his crew

Drogue →

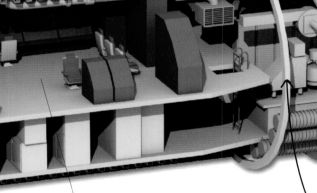

Oxygen tanks

Propeller

Prop drive shaft

Hydrogen tanks

Bulkhead

Control room The nerve center of the craft contains navigation equipment, sonar stations and communications. It also has controls for sailing depth, direction and speed, as well as weapons.

Fuel cells and diesel-engined generator The fuel cell system, with either nine smaller or two larger cells, is just in front of the diesel engine. The fuel cells combine oxygen and hydrogen to make water and generate an electric current.

Electric drive motor The Permasyn 1.7-MW (1.7 million watts) electric propulsion motor can switch between electricity from the fuel cells or the diesel generator.

The overall length of the 212 is 184 feet (56 m), with a beam (width) of 23 feet (7 m). Its weight is 1,600 tons (1,450 metric tons).

The first Type 212, known as the U-31, took to the water in 2002 and entered commission in 2005. Running with its quiet electric motor, it is very difficult to detect by passive sonar.

NIMITZ SUPERCARRIER

A floating, self-powered city designed for warfare, the supersized aircraft carrier is the world's largest mobile military base. These giant vessels are nuclear-powered, using uranium fuel pellets smaller than suitcases. If they had to carry liquid fuel for diesel or gas turbine engines, this would take up vast amounts of space inside and severely limit the craft's time at sea.

Eureka!

The first successful nuclear-powered ship was not an aircraft carrier but a submarine, USS *Nautilus*, launched in 1955. It was followed by the Russian icebreaker *Lenin* in 1957. In 1958, *Nautilus* sailed across the Arctic ice cap to the North Pole.

What next?

Replacing the Nimitz carriers will be the Gerald R. Ford class, of similar size but with advanced reactors and electronics. Building work started in 2007 and the first ship should be ready by 2016.

Island The bridge and other main control rooms are located here. Observers constantly watch aircraft, on radar and by eye, to avoid collisions.

The position of the island on the right of a carrier's deck dates back to when propeller-driven aircraft took off and landed. The direction of the prop's rotation meant these planes were more likely to veer to the left than the right.

Radar

Props

Reactors and propulsion Two A4W nuclear reactors are heavily shielded by thick metal to prevent radioactivity leaks. They boil water into steam to drive the four steam turbines that turn the propellers.

All HOOKED UP!

As well as having to accelerate very quickly at take-off, aircraft landing on a carrier must slow down, or decelerate, just as rapidly. The usual method is a long bar with a U-shaped claw at the end, the tailhook or arrestor hook, that lowers from under the aircraft's tail as it prepares to land. At touchdown this catches on one of several steel cables or wires strung across the deck. As the cable run outs, more and more braking force is applied to slow it down or arrest it, so the aircraft stops quickly but smoothly.

A fighter "catches a wire" on the deck of an aircraft carrier

With an all-up weight of more than 110,0 tons (100,000 metr tons), the Nimitz cla are the largest milita ships in the world. However the cruise li Oasis of the Seas launched in 2008, is 250,000 tons (225,00 metric tons).

Discover what life is like on a Nimitz supercarrier by visiting www.factsforprojects.com and clicking on the web link.

The Nimitz class carriers are named in honor of Fleet Admiral Chester Nimitz, commander in the US Pacific during World War II.

Jet blast deflector

F-18 Hornet fighter ready for take-off

Catapult shuttle engages in the aircraft's nose gear

Catapult control pod

Catapult officer

Flight deck crew

Catapult track set into deck

✳ How do STEAM CATAPULTS work?

Despite a supercarrier's size, it has a relatively short deck from which aircraft take off. So a steam-powered catapult system hidden under the deck boosts a plane's speed at the start of its take-off run. A bullet-shaped shuttle slides in a long hollow track in the deck, and attaches to a quick-release catch at the plane's nose wheel. A blast of high-pressure steam piped from the ship's boilers then pushes a cylinder that pulls along the shuttle, to accelerate the plane before its own engines take over.

Angled flight deck
The total flight deck length is more than 1,083 feet (330 meters). Planes can both take off and land within seconds of each other using the additional angled side deck.

The lead ship of the class, USS Nimitz, went into service in 1975.

Take-off ramp Aircraft need different take-off runs according to their type and their weapons and fuel loads. The angled bow ramp (at the front) gives them an added uplift.

The reactors of Nimitz carriers do not need refueling for 20 years. However, the refuelling process must be done with great caution and is only possible at certain ports.

Hangars Aircraft are folded up for storage on the main hangar deck, which is 682 feet (208 m) long, 108 feet (33 m) wide and 26 feet (8 m) high.

Accommodation More than 5,000 people take it in shifts to eat, relax, sleep and go on active duty.

The carrier has enough aircraft fuel for 13 days of non-stop action.

GLOSSARY

Alloy

A combination of metals, or metals and other substances, for special purposes such as great strength, extreme lightness, resistance to high temperatures, or all of these.

Antenna

Part of a communications system that sends out and/or receives radio signals, microwaves or similar waves and rays. Most antenna (also called aerials) are either long and thin like a wire or a whip, or dish-shaped like a bowl.

Barrel

The long, tube-shaped main part of a gun or similar weapon, in front of the breech and ending at the muzzle (open end).

Bearing

A part designed for efficient movement, to reduce friction and wear, for example, between a spinning shaft or axle and its frame.

Boom

A long, slim, arm-like part of a crane or similar machine that can usually move up and down, from side to side and perhaps in and out.

Bow

In watercraft, the forward-pointing part of the hull (main body).

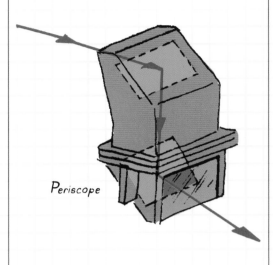
Periscope

Tiltrotor

Breech

The rear part of a gun or similar weapon, at the opposite end to the muzzle, where the bullets or similar ammunition are loaded.

Bridge

In watercraft, the control room of a large boat or ship, housing the wheel (steering wheel), engine throttles, instrument displays and other important equipment.

Bulkhead

An upright wall or partition across the width of a structure, such as across the hull of a ship or the fuselage of an aircraft.

Bullet

Usually a short, rod-shaped, solid metal or plastic object with a pointed front end, fired out of a gun or similar weapon.

Caliber

In weapons, the inside width of a tube such as a gun barrel.

Cartridge

A package loaded into a gun consisting of the bullet or shell, along with its casing, the main explosive and the primer.

Cylinder

In an engine or mechanical part, the chamber inside which a well-fitting piston moves.

Fuselage

The main body or central part of an aircraft, usually long and tube-shaped.

Gears

Toothed wheels or sprockets that fit or mesh together so that one turns the other. If they are connected by a chain or belt with holes where the teeth fit, they are generally called sprockets. Gears are used to change turning speed and force, for example, between an engine and the road wheels of a car, or to change the direction of rotation.

GPS

Global Positioning System, a network of more than 20 satellites in space going around the Earth. They send out radio signals about their position and the time, allowing people to find their location using GPS receivers.

Gyroscope

A device that maintains its position and resists being moved or tilted because of its movement energy, usually consisting of a fast-spinning ball or wheel.

Hull

The main body of a water vessel such as a ship, and also of some land vehicles such as tanks.

Hydraulic

Machinery that works by using high-pressure liquid such as oil or water.

Infrared

A form of energy, as rays or waves, which is similar to light but with longer waves that have a warming or heating effect.

Laser

A special high-energy form of light that is only one pure color, where all the waves are exactly the same length, and are parallel to each other rather than spreading out as in normal light.

Mine

A "bomb in the water"—a weapon floating in the water that explodes when it touches or comes near to another object. Some mines are buried on land.

Muzzle

The open end of a gun or similar weapon, where the bullet or other ammunition comes out.

Piston

A wide, rod-shaped part, similar in shape to a food or drink can, that moves along or up and down inside a close-fitting chamber, the cylinder.

Primer

A small amount of explosive that is set off by a spark or by being hit, which then makes the main explosive blow up and fire the bullet or shell.

Caterpillar tracks

Radar

A system that sends out radio waves that reflect off objects such as aircraft or ships, and detects echoes to find out their position.

Radioactive

Giving off or emitting certain kinds of rays and/or particles known as radiation, which can cause harm to people and other living things.

Rudder

The control surface of an aircraft or watercraft, usually on the upright fin or 'tail' of an aircraft or below the rear hull of a boat, that makes it steer left or right (yaw).

Satnav

Satellite navigation, finding your way and location using radio signals from the GPS (Global Positioning System) satellites in space.

Shell

A bullet-like object fired out of a gun or similar, which usually contains explosives that blow up when they hit the target.

Stealth technology

Designing an aircraft, ship, or similar object so it is difficult to detect by sight, sound, heat sensors, or radar equipment.

Stern

In watercraft, the rear or blunt end of the hull.

Suspension

Parts that allow the road wheels or tracks of a vehicle to move up and down separately from the driver and passengers, to smooth out bumps and dips in the road. Also, any similar system that gives a softer, more comfortable ride.

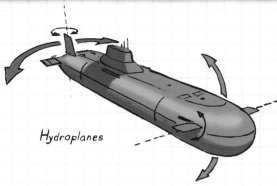

Hydroplanes

Tailplane

The two small rear wings on most aircraft and some helicopters, also known as the horizontal stabilizers. They carry the elevators and are usually next to the fin or "tail."

Thermal

To do with heat energy and temperature.

Torpedo

A self-powered exploding weapon launched at the water's surface or below, usually to hit a watercraft such as a ship, boat or submarine.

Traverse

When a gun or similar weapon aims by moving from side to side (horizontally) rather than up and down (vertically).

Turbine

A set of angled fan-like blades on a spinning shaft, used in many areas of engineering, from pumps and cars to jet engines.

Turbofan

A jet engine with fan-like turbine blades inside, and one very large turbine or "fan" at the front that works partly as a propeller.

Turboshaft

A jet engine with fan-like turbine blades inside, which spins a shaft for power rather than using its jet blast of gases.

INDEX

A-10 Thunderbolt **20–21**
AH-64 Apache helicopter **18–19**
active camouflage 24
active sonar 30
active stealth 24
AESA (Active Electronically Scanned Array) 24
afterburners 24, 25
ailerons 25
Airborne Warning and Control Systems (AWACS) 26
aircraft 6, 10, 19, 20, 21, 22, 23, 24, 26, 27, 28, 29, 30, 32, 34, 36, 37
aircraft carriers 32, 36
airframes 23, 25, 26
air intakes 11, 24
alloys 14, 29, 31
aluminum 14, 29
ammunition 9, 14, 16, 20, 21
anti-aircraft guns 20
anti-missile missiles 10
anti-radar coatings 32
armor 7, 12, 13, 14, 16, 20
Avenger minehunter **30–31**

B-52 Stratofortress **28–29**
bomber aircraft 6, 7, 9, 28
bomblets 15
bombs 10, 21, 28, 30, 34
Browning, John 8
bulkheads 29, 32, 35
bullets 8, 9

CAS (Close Air Support) 20
caterpillar tracks 12
chobham armour 17
Cold War 28
computers 10, 22, 24, 26, 30, 31
control surfaces 6, 34
cooling systems 33
crew 14, 16, 18, 22, 26, 28, 33, 35, 37
cruise missiles 10, 11
Cummins VTA-903T turbo diesel engine 12

Daring, HMS 32
delta wings 25
destroyers 32, 33
detonations 16
diesel engines 13, 31, 34, 35, 36
dissipators 9
DM2A4 torpedo 34
drag 25
drive shafts 12, 19, 22, 32
drive wheels 12, 13, 14

E-3 Sentry Awacs **26–27**
electricity 25, 26, 32, 34, 35
electricity generators 25, 32

electric motors 15, 16, 21, 31, 32, 34, 35
electronics 7, 26, 36
elevators 24
engines 11, 12, 13, 16, 19, 20, 21, 22, 23, 24, 25, 27, 28, 29, 30, 31, 32, 37
exhausts 25, 32, 33
explosives 7, 8, 11, 20, 28

F-18 Hornet fighter 37
F-22 Raptor **24–25**, 32
F-35 Lightning 20, 24
F-86 Sabre 9
fighter aircraft 6, 7, 9
flight decks 22, 26, 37
fly-by-wire 24
fuel 10, 16, 22, 23, 24, 25, 26, 28, 34, 36
fuel cells 35
fuel loads 28, 37
fuel tanks 19, 29
fuselage 19, 20, 22, 23, 27, 29

gas turbines 32, 33, 36
Gatling gun 8, 20, 21
Gatling, Richard 20
GAU/8A Avenger Gatling gun 20, 21
gears 22, 32
generators 13, 25, 32, 35
glass-fiber plastic 30
GPS (Global Positioning Satellite) 10, 26
Granit missile **10–11**
grenades 14, 17
ground-mapping radar 26
ground vehicles 20
guided missiles 14
gunners 13, 14
gunpowder 6, 8, 14
guns 6, 7, 12, 13, 17, 18, 20, 33
gyros 10
gyroscope effect 9

Harpoon missiles 33
helicopters 9, 18, 19, 22, 23
Hellfire missiles 19
HIMARS (high Mobility Artillery Rocket System) 14
Honeywell AGT 1500C turbine engine 16
HTR (Hybrid Tandem Rotor) 22
Hydra rockets 19
hydraulics 15, 21
hydrogen 34, 35
hydrophones 31
hydroplanes 6, 34

IFV (Infantry Fighting Vehicle) 12
infrared (heat) rays 16
infrared signatures 32

ironclads 32

jet engines 10, 11, 20, 21, 24, 25, 26, 34

landing gear 26
launcher modules 15
launcher turrets 15
launch tubes 10, 11
lifeboats 33
lookdown radar 27

M1 Abrams tank **16–17**
M2 .50 Caliber heavy machine gun 8, 9
M2 Bradley IFV **12–13**, 14, 15
M26 rockets 14, 15
M230 chain gun 18
M242 Bushmaster chain gun 12
M269 launcher 15
M270 rocket launcher **14–15**
M3M 50-cal machine gun **8–9**
M61 Vulcan rotary cannon 29
machine guns 8, 9, 13, 17
magnetic detectors 35
Maverick missiles 21
microwaves 7, 26, 30
mid-air refueling 28, 29
mines 28, 30
missile pods 15
missiles 6, 7, 10, 11, 14, 18, 19, 20, 21, 24, 26, 28, 32, 33, 34
munitions 8, 10, 11

navigation systems 26, 35
night vision 13, 16
Nimitz, Chester, Field Admiral 37
Nimitz supercarrier **36–37**
nuclear power 34, 36
nuclear reactors 34, 36
nuclear weapons 7, 11, 28

outrigger wheels 29
oxygen 35

passive sonar 31, 35
periscopes 16, 17, 35
Permasyn 1.7-MW electric propulsion motor 35
petrol engines 34
pilots 9, 18, 20, 22, 24, 28
pitch 6, 34
pneumatic mechanisms 21
Pratt & Whitney F119-PW-100 turbofan engine 25
Pratt & Whitney TF33-PW-100A turbofan jet engine 27
Pratt & Whitney TF33-P-3/103 turbofan jet engines 29

primer 8
probes 27
prop drive shafts 35
propellant 8
propellers 21, 22, 31, 32, 35, 36
proprotors 23
pulse doppler radar 27
pulse-jet engines 10

radar 7, 10, 24, 26, 27, 30, 31, 32, 33
radio 18, 24, 26, 27, 30, 32, 33, 36
radioactivity 36
radome 18
reactive armour 16
recoil mechanism 8, 20
refueling 23, 26, 34, 37
remote controlled aircraft 18, 28
RIBs (Rigid Inflatable Boats) 31
rocket boosters 11
rockets 14, 15, 18, 19, 26, 28
Rolls Royce TT406/AE 1107 turboshaft engine 23
Rolls Royce WR21 gas turbine 32
rotary cannons 7, 20, 21, 29
rotodomes 27
rotor blades 18, 19, 22, 23
rotor hubs 23
rotors 22, 23
ROVs (remote-operated vehicles) 31
rudders 6, 21, 24, 34

satellite systems 26
satnav 26
Sea Viper missiles 33
shells 7, 8, 12, 16, 17, 20, 21
ships 6, 7, 10, 17, 23, 26, 27, 30, 32, 33, 34, 37
Sidewinder missiles 19, 21
Sikorsky X/R-4 18
smoke grenade launchers 13, 17
smoothbore 17
solid fuel 11
sonar (Sound Navigation and Ranging) 30, 31, 34, 35
sonar arrays 30
sonar pods 30, 31
sonar signatures 34
speed of sound 10, 25
stealth 7, 24, 32
steam catapults 37
Stinger anti-aircraft missiles 13, 19
stub pylons 20
stub wings 19, 22
submarines 6, 10, 11, 32, 34, 35, 36
submunitions 14

supercarriers 37
suspension 13

T700-GE-701 turboshaft engine 19
tail fins 10
tail guns 29
tailhooks 36
tailplanes 11, 19, 20, 21, 25
tail rotors 19
take-off 24, 25, 29, 36, 37
tanks 6, 7, 9, 12, 13, 16, 2
tank transporters 17
telescopic loading booms 15
TF34-GE-100 turbofan jet engines 21
thrust 19, 24, 25
thrust vectoring 25
tiltrotors 22
tiltwings 22
Titanic 30
titanium 20
TNT (trinitrotoluene) 11
Tomahawk missile **10–11**
torpedoes 34
torque 19
troop carriers 12
troops 13, 20, 22
turbines 25, 32, 36
turbojet engines 11, 25, 3
turrets 13, 15, 16
twin engine pods 28
Type 45 destroyer **32–33**
Type 212 submarine **34–3**

underwing pylons 29
uranium 36

V-1 Flying Bomb 10
V-22 Osprey Tiltrotor **22–23**
vertical take-off or landing 22
Vietnam War 17
vision blocks 16

warheads 11, 14
Watson-Watt, Robert 26
Waukesha MCM diesel engine 31
Williams F107 turbofan jet engine 11
wings 10, 11, 20, 21, 22, 23, 27, 28, 29
wingtip tanks 29
World War I 6, 8, 16, 28
World War II 6, 10, 12, 18, 21, 26, 37

yaw 34